SCL
En 33
2011
1-5
v.2

T5-AQQ-815

Aisha Makes Work Easier

An Industrial Engineering Story

Written by the Engineering is Elementary Team

Illustrated by Jeannette Martin

T70059

Education Resource Center
University of Delaware
Newark, DE 19716-2940

Chapter One | Breakfast with Malcolm

"Breakfast is served!" Aisha danced across the kitchen with a carton of milk in her hand.

"It won't be served if you spill that milk all over the floor," her brother Malcolm said.

Aisha grinned. Malcolm shook his head and smiled back. Aisha and Malcolm didn't get to hang out as much as they used to, now that he was in college studying to be an industrial engineer. But today their dad had to work at his restaurant, and he was leaving Malcolm to look after Aisha and their cousin Tanya.

When Malcolm first went to college, Aisha had no idea what an engineer did. But now she knew: engineers are people who use their knowledge of math and science to design things or processes that solve problems. Aisha was very proud of her brother.

Aisha's dad walked into the kitchen. "So, how are my two favorite children going to spend their Sunday in Boston?" he asked.

Aisha planted her hands on her hips. "Tanya and I have big plans for you today, Malcolm."

Malcolm raised his eyebrows. "Oh, really?"

Aisha nodded. "You can take Tanya and me to work with you!" Malcolm had gotten a summer job at a nearby factory—a potato chip factory! Aisha had been begging Malcolm for weeks to take her to work with him. "We'll go to the potato chip factory," Aisha continued, "and I'll make up my own flavor and they'll name it after me and then I'll be famous and—"

"There she goes again," Dad said.

"Whoa there, kiddo," said Malcolm. "In case you've forgotten, today is Sunday. The potato chip factory is closed. That's why I'm lucky enough to be your babysitter, remember? Besides, I told you we could go only if you and Tanya had completed your summer projects."

"Oh, yeah," Aisha sighed. It was almost the end of summer, and Aisha and Tanya were supposed to have projects to show on the first day of school. The projects had to be about something they'd learned over their vacation. Aisha couldn't understand why she had to do any schoolwork at all in the summer.

"Maybe you and Tanya can come up with a project in the morning, and then we could go do something in the afternoon," Malcolm said.

Dad gathered his uniform and work bag from the table. "Now that you have a plan, I'm off to the restaurant to create my latest masterpiece: the daily special! I'll see you all for lunch."

| Chapter Two | # Malcolm's Game |

When Tanya arrived, the girls sat in the dining room, trying to think of something they'd learned or done over the summer that would make a good project. Malcolm sat down at the other end of the table, spreading out some drawings.

"Since you two are finally getting down to business, I thought I'd do some work of my own," he said.

Tanya peered over his shoulder. "I thought you worked at a potato chip factory. What do these drawings have to do with making chips?"

"Or eating them?" Aisha asked. "You do get to eat the chips, don't you?"

"Sometimes," said Malcolm, "but mostly I help the factory improve the systems that their workers use. I try to make their work easier. Or faster or safer. That's what industrial engineering is all about."

"What do you mean?" Tanya asked.

"I'll give you an example. See my textbooks?" He pointed to a pile of thick books. "Pretend they're a crate of potatoes. Try pushing them across the table."

Aisha pushed on the books and slid them from one end of the table to the other. "Wow, they're pretty heavy."

Malcolm reached into his book bag and pulled out a bunch of pencils. He lined them up on the table and put the pile of books on top. "Try pushing them now."

"Hey, that's a lot easier!" Aisha exclaimed.

"The pencils act like little wheels—they help make work easier by reducing friction," Malcolm explained.

"Making work easier!" Tanya said. "Last year in school we studied things that help make work easier—simple machines. Things like levers and wheels and axles, right?"

"Exactly," said Malcolm. "We use lots of simple machines at the factory. All the big machines are made up of smaller parts, called subsystems. A lot of the time those subsystems are made up of simple machines. The simple machines help the workers get more done, because each job is easier. And the workers are less likely to get hurt, too."

"That's what we can do our project on!" Aisha cried. "We can show how simple machines make the work at your factory easier."

"Now you're thinking," Malcolm said. "Hmm, with that project it looks like you'll finally get your wish to go check out the chips. We'll go sometime this week." The girls cheered. "But," he added, "I think there's a lot of cool stuff you could learn about simple machines before we head to the factory."

Aisha frowned. "Like what?"

"Simple machines aren't just used in factory subsystems," Malcolm said. "They make work easier all over. Let's go on a scavenger hunt. If you show me some simple machines that you spot today, I'll show you how I use them at the factory. Deal?"

"Deal!" the girls shouted.

"I think I know a good place to start," said Malcolm. "Let's go visit my friend Sean at the theater. I bet he'll have lots of simple machines to show us."

As Aisha reached for the door of the coat closet, she called out, "Hey! I found one!" She pointed to the door handle. "A lever."

"You're right," said Malcolm. "And inside the handle, where we can't see, there's another simple machine: a wheel and axle. See, we haven't even left the house yet, and you're already well on your way."

Chapter Three | Simple Machines on Stage

Aisha, Tanya, and Malcolm stepped into the theater where Sean worked.

"Hi," Sean called as he walked into the empty lobby to greet them. The matinee didn't start for a few hours.

"It's so neat that you get to hang out here, Sean," said Aisha. "This building is so beautiful."

"Yes, it is," Sean said. "But I do more than hang out here. I help set up the stage for some of the performances."

"If I worked here, all I would want to do is dance!" Aisha said. She twirled around to demonstrate. She and Tanya started leaping and skipping around the lobby.

"Come on, Twinkle Toes," Sean called. "Let me show you the rest of the theater." He led them down the side aisle, onto the stage, and into the wings. The huge wooden floor gleamed under the lights. Aisha imagined herself dancing across the stage in front of hundreds—no, thousands—of people in the audience.

"Let me show you how we change scenes during a performance." Sean walked over to a control panel full of buttons. He pressed one, and without a sound, an outdoor scene with trees, plants, and a park bench floated down from the ceiling.

"Wow! It's like magic," Aisha exclaimed.

Sean pointed to the machinery hanging from the ceiling.

"It's actually just a bunch of pulleys—wheels with ropes that run around them."

"Pulleys!" shouted Aisha and Tanya. Malcolm gave them a thumbs-up sign.

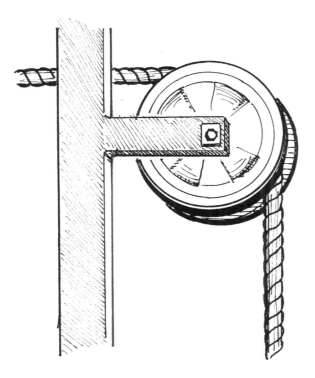

"Am I missing something here?" Sean asked.

Tanya explained the simple machines scavenger hunt to Sean.

"I see," said Sean. "Well, we use pulleys all the time around here to change the direction of forces. I pull down on one end of the rope and, thanks to the pulley, the other end of the rope pulls the scenery up."

"Imagine how much harder it would be for the stage crew to get all that scenery into place without help from the pulleys," Malcolm said.

"The simple machines behind the magic," said Aisha.

"Exactly," said Malcolm. "You girls are naturals at this game. Keep it up and you could be engineers!"

"Engineers who dance!" Aisha said, leaping across the stage. Joining in, Tanya improvised a few steps. They imagined themselves in beautiful costumes, fans showering them with thunderous applause.

"Now that you girls have danced up a storm, you must be getting hungry," Malcolm said. "Let's go see what Dad's cooking up for lunch."

Chapter Four | Simple Machines on the Streets

On the way to lunch, Malcolm, Aisha, and Tanya walked through Boston Common towards the State House. Aisha looked up at the stone monument towering high above her. The writing on the monument said it was dedicated to Colonel Robert Shaw and the 54th Regiment of the Union Army. They were the first all-black volunteer regiment to fight in the Civil War. She began a marching dance to remember them.

"This is one of my favorite neighborhoods," said Tanya. "The houses and apartments are so old, I can imagine horse-drawn carriages on the streets."

"The cars around here now are a lot fancier than carriages," said Aisha. She paused for a few moments,

and then said, "You know what? Maybe they're not that different after all. Cars and carriages both work because of their wheels and axles."

"That's a great observation," said Malcolm. "Once you start spotting simple machines, you can start to see how complicated systems, like cars, are just arrangements and combinations of simple machines."

"Aisha's not the only one who can spot simple machines!" exclaimed Tanya. "Check out that flagpole on the schoolhouse. There's a pulley on it!"

Aisha looked up at the American flag flapping in the warm summer breeze. Tanya was right. There was a pulley, helping to mark an important Boston site—the Abiel Smith School and Museum of Afro-American History. The Smith

School was the first public school for African-Americans in the whole country.

"Come on, girls," said Malcolm. "I bet Dad has lunch on the table. I'm sure he'll want to hear about the progress you've made on your project." Aisha and Tanya took off running towards Dad's restaurant with Malcolm close behind.

Delicious smells greeted Malcolm and the girls as they walked into the restaurant. They were careful to avoid the cooks and other workers in the kitchen as they made their way to the small prep area in the back, where Dad was waiting for them.

"I hope you're hungry," Dad said. "I've started cooking lunch and could use your help."

Malcolm took out dishes and utensils, while Tanya and Aisha told Dad about the simple machines they'd seen so far. Dad picked up a knife and began cutting dough to make rolls.

"Dad," Malcolm said, winking, "nice technology you have there. Is that a wedge you're using?"

"That's exactly what it is," said Dad. "Did you girls know that a wedge is a simple machine that helps push things apart? The thin edge of the knife concentrates the force I put on it to push the dough apart."

"Malcolm, I know the knife is a simple machine, but what did you mean when you said it's technology?" Aisha asked. "It's not electric or anything."

"Technology is more than just electronic gadgets," said Malcolm. "Technology is any thing or process that people design to solve a problem."

"Oh," said Tanya. "So, that knife—I mean, the wedge—helps you solve the problem of cutting stuff."

"Yes," said Dad. "This knife makes my life as a chef much easier. Why don't you grab a few different wedges and help finish our lunch? Tanya, you can make some lemonade, and Aisha, you can cut out some cookies."

Tanya picked up the juicer that was sitting on the counter. It was the old-fashioned kind with a cone sticking out of a cup to catch the lemon juice. "So this is a wedge, too," she said. "The pointy part pushes into the lemon to squeeze out the juice."

"Aisha, how are you doing with those cookies?" Dad asked.

"This cookie cutter is another helpful wedge," said Aisha. "It helps me cut perfect star shapes."

Malcolm put the cookies in the oven to bake, and Dad put sugar in the lemonade Tanya had squeezed. While Dad finished grilling chicken, Tanya and Aisha looked around

the rest of the kitchen. They found gears in an old-fashioned eggbeater, levers and a screw in a wine bottle opener, and all kinds of wedges for cutting.

Finally, lunch was ready. "Now that you two have been using both your brains and your hands, I bet you've worked up quite an appetite!" Dad said.

"You better eat up," said Malcolm. "I have one more special stop planned for us this afternoon."

Chapter Five | A Visit to the Museum of Science

After lunch, Malcolm led the girls to the Museum of Science. As soon as she stepped inside, Aisha's eyes lit up.

"What is that?" asked Aisha, pointing to the machine in front of her. The machine, inside a plastic case, had lots of metal parts that were moving some small balls around.

"I bet we can find some simple machines in there," Tanya said.

"This is like a Rube Goldberg machine," Malcolm explained. "Usually when someone designs a machine, they try to make a complex process simple, using as few steps as possible. The goal of Rube Goldberg machines is just the opposite.

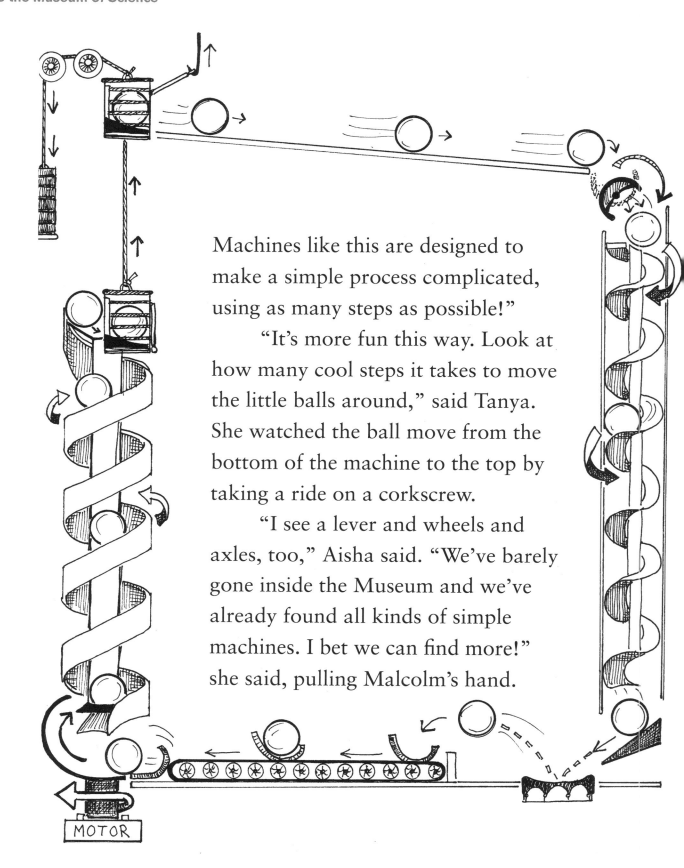

Machines like this are designed to make a simple process complicated, using as many steps as possible!"

"It's more fun this way. Look at how many cool steps it takes to move the little balls around," said Tanya. She watched the ball move from the bottom of the machine to the top by taking a ride on a corkscrew.

"I see a lever and wheels and axles, too," Aisha said. "We've barely gone inside the Museum and we've already found all kinds of simple machines. I bet we can find more!" she said, pulling Malcolm's hand.

As they walked up to the huge model of the *Tyrannosaurus rex*, Tanya pointed out the sharp, pointy wedges that he had for teeth.

"I wonder if dinosaurs could dance." Aisha giggled. "I think the *Tyrannosaurus rex* would dance like this." She puffed up her chest and rocked back and forth with big, lumbering steps as Tanya and Malcolm laughed.

They moved on to the baby chick incubator where several chicks were about to hatch.

"So, Aisha, you were inspired by the dinosaur," said Malcolm. "Do the little chicks give you dance fever too?"

"Actually, they're awesome!" exclaimed Aisha. "The chicks hatching would make a great modern dance. First I would start like an egg," she said, crouching down into a little ball. "And then I would use my beak like a wedge to help me break out of my shell," she continued. She flapped her arms and ran in a small circle.

Tanya and Malcolm laughed at the sight of Aisha making scratching and pecking motions. She did look like a little chick!

"Bravo!" exclaimed Malcolm. "Why don't we dance our way over to the exhibit I want to see?"

When they entered the Science in the Park exhibit,

Aisha and Tanya saw children playing on merry-go-rounds and racing toy cars down inclined planes. Malcolm led them to the seesaw.

"Hey," Aisha said. "That seesaw is a simple machine, isn't it?"

"You're absolutely right," said Malcolm. "But which simple machine is it? How do you think it works?"

"That's easy," said Tanya. "It's a lever—a board or a bar."

"Don't forget the fulcrum!" said Aisha, pointing to the support under the bar. "That's what lets the seesaw swing back and forth," She stretched both of her arms out and swayed like the seesaw.

"Hey, Malcolm," said Tanya, "I bet Aisha and I can lift you up on the seesaw."

"All right, let's see what you can do," Malcolm said. He climbed onto one end of the seesaw, and both girls climbed on the other. Malcolm's end of the board rose and carried him up.

"That was too easy," Malcolm said. "I've got another puzzle for you. Can just *one* of you lift me up?"

Everyone got off the seesaw. Aisha and Tanya looked at Malcolm, then the seesaw, then each other.

"Hmm," said Tanya. "I bet I can do it." She climbed on

the seesaw close to the pivot point. "Now you get on," she told Malcolm. He sat on the other end. His weight pushed Tanya way up in the air.

"Whoa!" Tanya said. "I guess I was wrong about what would happen when I moved towards the fulcrum."

"There's an important clue in that mistake," said Malcolm. "Now that you've seen what happens, do you have any new ideas?"

"I do," exclaimed Aisha. "If Malcolm moves to the middle and Tanya moves back to the end, I bet it will be a lot easier to lift him."

Sure enough, Tanya's weight raised Malcolm high in the air.

"You got it! See? Changing the distance between the fulcrum and the weight changes the force you need to use to lift it. OK, you can let me down now," Malcolm said. But Tanya only giggled and stayed in place.

"C'mon, Tanya!"

"Nope!" she teased.

"Well, if you don't let me down, then we can't leave. And if we don't leave, then I can't take you to—"

"The potato chip factory!" cried Aisha. Tanya quickly slid off of the seesaw, and the three of them made their way back home.

Chapter Six | Fun at the Factory

The next day, Aisha and Tanya skipped into the potato chip factory ahead of Malcolm. His boss, Ms. Johnson, was there to meet them.

"Hello, girls!" Ms. Johnson said. "We're excited to have you here. Malcolm told me you know a lot about the simple machines we use in our factory subsystems."

"We found simple machines everywhere we went yesterday," said Aisha. "I really want to see how they work here in the factory. I know it will help us with our summer project for school."

Ms. Johnson and Malcolm led the girls to a big truck. "This is the beginning of the whole process," she said. "There are about fifty thousand pounds of potatoes in here."

"Fifty thousand pounds!" cried Tanya. "That's a lot of chips."

The girls watched a worker drive a tractor up to the back of the truck, pick up a load of potatoes, and dump them on a conveyor belt. The conveyor belt didn't look like the ones at the supermarket checkout. There were lots of little bumps on it that bounced the potatoes around.

"Look!" Aisha said. "The conveyor belt uses wheels and axles to bounce all the dirt off the potatoes."

"That's right. Next we clean the potatoes even more," Ms. Johnson explained. "The potatoes go into a big vat of water where we wash them. On the other side of the vat, a big screw, called an auger, pulls the potatoes out again."

"At the Museum of Science we saw a screw lifting things, too," said Tanya. The group walked on.

"I see another simple machine!" cried Aisha. She pointed towards the machine that peeled the potatoes.

"All the potatoes slide down an inclined plane!"

"So the potatoes all come down gently," said Tanya.

"That's right," said Ms. Johnson. "We don't want mashed potatoes. We do want them sliced, though." She pointed to a round machine. "The potatoes are spun against really sharp blades—"

"And blades are wedges!" Aisha interrupted.

"Right! After the potatoes are sliced, they fall onto a conveyor belt. Then they're ready to be cooked in hot oil," said Ms. Johnson. The girls watched the potato slices fall into a huge kettle of oil.

Malcolm pointed towards metal rods that were stirring the kettle. "That mixing machine stirs the slices so a person doesn't have to stand over the hot oil doing that job."

"I bet an industrial engineer figured that out. Making work easier," Aisha said.

"After the potatoes cook for about five minutes, we add seasonings," said Malcolm.

"Like vinegar and herbs and barbecue and pepper," said Aisha.

"Not all at once, please," Malcolm joked.

"Then the chips are packaged and sent all over the country," said Ms. Johnson. "But we keep some bags for ourselves. How would you like some fresh potato chips?"

"You don't need to ask me twice," said Aisha. "Lead the way!"

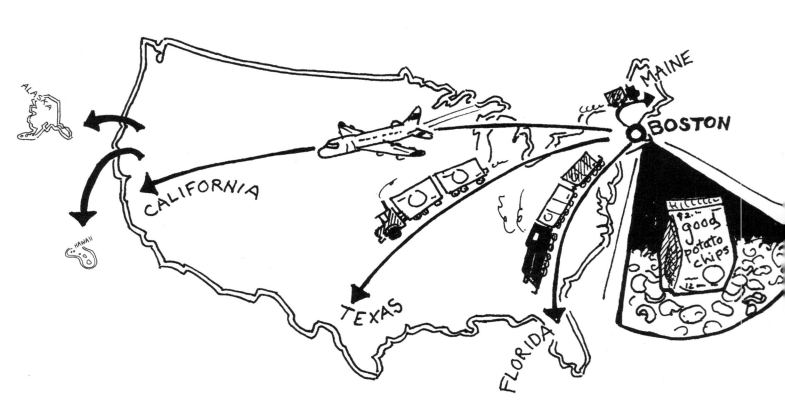

Chapter Seven | A Potato Chip Treat

At home that afternoon, Tanya and Aisha talked about how to present their project in class.

"We could write about it," said Tanya. "But I'd rather just take everybody in our class to the factory."

"What if we bring the factory to the class?" Aisha asked. "We could design our own subsystem just like they have at the factory and show it to our class!"

"That's a great idea," said Tanya. "We can pretend a truck drops off potatoes on the floor. If you have to keep reaching down to the floor so you can cook them, that will hurt your back. We can design a simple machine subsystem to lift the potatoes up!"

"Let's get Malcolm to help us," said Aisha.

The girls ran to find Malcolm and tell him their idea. Malcolm helped the girls gather wooden dowels, tubes, and long sheets of cardboard.

"Okay, now what?" Tanya asked.

But Aisha had already started pulling materials out of the box and putting things together. "Just dive in and start building stuff!" she cried.

A Potato Chip Treat

Malcolm interrupted. "Aisha, that's one way to get things done, but engineers usually make a plan first. It can be really helpful."

"But how am I supposed to know what to plan?" asked Aisha. "I figure things out as I go along."

"Sometimes engineers do that," explained Malcolm, "but we have steps to help us, too, to be sure that we know what materials we'll need and how they'll fit together. It's all part of the engineering design process."

"What's that?" asked Tanya.

"It's a system—a way of doing things, step by step," said Malcom. "You've already started asking some good questions to get yourself started. Next you need to imagine some solutions, and then plan it out. You may even want to draw out your plan."

Aisha frowned. "I can definitely imagine lots of things, but planning is no fun."

"Come on, Aisha," Tanya said. "We'll do it together."

Aisha and Tanya sat down on the floor and began sketching. They questioned, compromised, and built on one another's ideas until they had a plan that they thought would work.

"I think we're ready to go," Tanya told Malcolm.

"Let's take a look at your plan," Malcolm said.

41

The girls explained their drawing to Malcolm.

Malcolm nodded.

"Looks like you're ready for the next step—creating your subsystem!"

All three of them began working to bring the plan to life, testing each part along the way.

"I think we're done," said Tanya as she taped a pulley to the edge of the counter.

"How will this work?" asked Malcolm.

"I was thinking it would work kind of like the seesaw we used in the museum," said Tanya. "The potatoes will go on one end and then some weight at the other end will push the seesaw down. Then the potato will roll down the ramp onto the counter."

"Hmm," said Malcolm. "Let's give it a try."

Aisha pushed down on one end of the seesaw. The three of them watched as a few roly poly potatoes teetered off the side of the ramp. "Oh no!" Aisha cried. "Since there aren't any sides on this inclined plane, the potatoes fall off. That's no good!"

Malcolm looked at the girls. "It looks like it's time for

the last step of the engineering design process—improve. Do you think another simple machine might do the job better?"

"Well," said Tanya, "we could use an inclined plane like the one we saw at the factory. Maybe that would be better."

"And we could even make a cart with wheels and axles to carry the potatoes up the inclined plane!" Aisha added.

The girls and Malcolm worked on their inclined plane and cart until they were satisfied with the whole system.

"Well, look at that," said Malcolm. "You girls really are engineers!"

"I can't wait to show everyone at school all the stuff we learned this summer," said Aisha.

The girls began to giggle as they danced their way over to Malcolm to give him a big hug.

Try It!
Design Your Own Simple Machine System

Take a simple machines scavenger hunt of your own. Look around your house and neighborhood. What simple machines can you find? Your goal is to design a simple machine system that makes work easier.

Materials

- ☐ Potatoes, toys, or other weights
- ☐ Paper cups
- ☐ String or fishing line
- ☐ Chopsticks or sticks from outside
- ☐ Broomstick or yardstick
- ☐ Rubber bands
- ☐ Empty paper roll or spool
- ☐ Cardboard
- ☐ Tape

Simple Machines at Work

How can simple machines help you with the work you do every day? Do you ever need to lift your toys up so you can put them away? Have you ever tried to move a heavy box or bookbag across a room? You can design a simple machine system to help you with this work. Think about the problem you are trying to solve. Which simple machines might be best to use?

Build Your Simple Machines

With the suggested materials and with other things you might find around your house or yard, can you design and create some simple machines? Could a broomstick help you create a lever? Could an empty paper roll be made into a pulley? Try to lift a bag containing a few potatoes, toys, or weights using each of the simple machines you made.

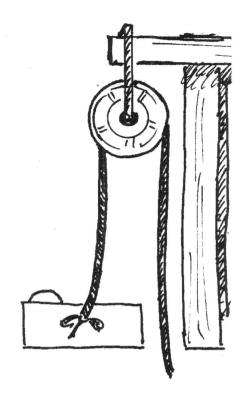

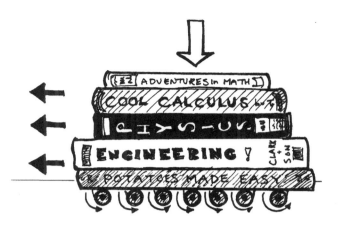

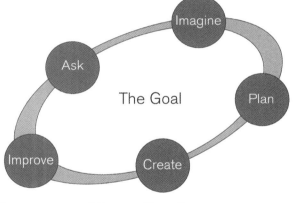

The Goal

Ask
Imagine
Plan
Create
Improve

Design Your System

Combine the simple machines you created to form a system. Begin by drawing a picture of what your system will look like. Then put your system together and test it. Does your system make moving your object easier?

Improve Your System

Use the engineering design process to improve your simple machines system. Go to the library and learn more about simple machines and how they work. Think about other materials you could use. Can you make a better system?

See What Others Have Done

See what other kids have done at http://www.mos.org/eie/tryit. What did you try? You can submit your solutions and pictures to our website, and maybe we'll post your submission!

Glossary

Engineer: A person who uses his or her creativity and understanding of mathematics and science to design things that solve problems.

Engineering Design Process: The steps that engineers use to design something to solve a problem.

Inclined Plane: A flat surface set at an angle that can be used to raise an object.

Industrial Engineering: The branch of engineering concerned with improving manufacturing or commercial systems and making work easier, faster, and safer.

Lever: A simple machine made up of a bar that pivots on another point, called a fulcrum.

Process: A series of actions or steps leading to a result or goal.

Pulley: A simple machine made up of a grooved wheel that is used to change the direction of force.

Screw: A rod with spiral threads.

Simple Machine: A mechanical device that changes the direction or magnitude of a force. Simple machines only require the application of a single force to work. The traditional list of simple machines is: inclined plane, screw, wedge, lever, wheel and axle, and pulley.

Subsystem: A system that is part of some larger system.

System: A group of parts that work together to meet a goal.

Technology: Any thing or process that people create and use to solve a problem.

Wedge: A simple machine made up of a piece of material tapered to a thin edge on one side.

Wheel and Axle: A simple machine made up of two circular or cylindrical components of different sizes (radii). The wheel is the larger circle or cylinder and it turns the attached, smaller circular/cylindrical part, the axle.

Work: The amount of effort it takes to accomplish something.